Joseph Abruscato   Joan Wade Fossaceca   Jack Hassard   Donald Peck

HOLT SCIENCE

Holt, Rinehart and Winston, Publishers
New York · Toronto · Mexico City · London · Sydney · Tokyo

THE AUTHORS

Joseph Abruscato
Associate Dean
College of Education and Social Services
University of Vermont
Burlington, Vermont

Joan Wade Fossaceca
Teacher
Pointview Elementary School
Westerville City Schools
Westerville, Ohio

Jack Hassard
Professor
College of Education
Georgia State University
Atlanta, Georgia

Donald Peck
Supervisor of Science
Woodbridge Township School District
Woodbridge, New Jersey

Cover photos, front and back: Z. Leszczynski/Animals Animals.
The animals on the front and back covers are baby raccoons. Raccoons are
mammals that can be found over most of North America. They eat roots, berries,
fish, and other small animals. Raccoons often live in suburban areas, finding
food in garbage cans and gardens.

Photo and art credits on page 168

ACKNOWLEDGMENTS

Teacher Consultants

Armand Alvarez
District Science Curriculum Specialist
San Antonio Independent School District
San Antonio, Texas

Sister de Montfort Babb, I.H.M.
Earth Science Teacher
Maria Regina High School
Uniondale, New York
Instructor
Hofstra University
Hempstead, New York

Ernest Bibby
Science Consultant
Granville County Board of Education
Oxford, North Carolina

Linda C. Cardwell
Teacher
Dickinson Elementary School
Grand Prairie, Texas

Betty Eagle
Teacher
Englewood Cliffs Upper School
Englewood Cliffs, New Jersey

James A. Harris
Principal
Rothschild Elementary School
Rothschild, Wisconsin

Rachel P. Keziah
Instructional Supervisor
New Hanover County Schools
Wilmington, North Carolina

J. Peter O'Neil
Science Teacher
Waunakee Junior High School
Waunakee, Wisconsin

Raymond E. Sanders, Jr.
Assistant Science Supervisor
Calcasieu Parish Schools
Lake Charles, Louisiana

Content Consultants

John B. Jenkins
Professor of Biology
Swarthmore College
Swarthmore, Pennsylvania

Mark M. Payne, O.S.B.
Physics Teacher
St. Benedict's Preparatory School
Newark, New Jersey

Robert W. Ridky, Ph.D.
Professor of Geology
University of Maryland
College Park, Maryland

Safety Consultant

Franklin D. Kizer
Executive Secretary
Council of State Science Supervisors, Inc.
Lancaster, Virginia

Readability Consultant

Jane Kita Cooke
Assistant Professor of Education
College of New Rochelle
New Rochelle, New York

Curriculum Consultant

Lowell J. Bethel
Associate Professor, Science Education
Director, Office of Student Field Experiences
The University of Texas at Austin
Austin, Texas

Special Education Consultant

Joan Baltman
Special Education Program Coordinator
P.S. 188 Elementary School
Bronx, New York

CONTENTS

CHAPTER 1

USE YOUR SENSES

1.

SENSES

You find out many things.
You use your **senses**.
Your eyes help you **see**.
Your hands help you **feel**.

Your ears help you **hear**.
What do they hear?

Use your nose.
You can **smell**.
You can **taste**.
What do you use?
How do the children use
their senses?

ACTIVITY

What kind of food is it?

1. Find out about the food.

2. Use your senses.

3. What do you see?
   What do you hear?
   What do you smell?
   What do you feel?
   What do you taste?

2.

SAME OR DIFFERENT?

The girls are twins.
They look the **same**.
They have **different** hats on.

You see signs every day.
Look at the colors.
Look at the shapes.
How are they the same?
How are they different?

How are the flowers the same?
How are they different?
The children use their senses.
They look, smell, and touch.

3.

GROUPING THINGS

You can **group** things.
Which foods taste sweet?
The ones that are the same go together.
They make a group.

You look and feel.
Some shells are the same.
Some shells are different.
You can put them into groups.

ACTIVITY

Can you group seeds?

1. Look at seeds.

2. Some are the same.
   Some are different.

3. Group them by colors.

4. Group them by shapes.

5. Group them by sizes.

PEOPLE AND SCIENCE

Firefighters use their senses.
They can see the fire.
They can feel the heat.
They can smell the smoke.
What can a firefighter hear?

Main Ideas

- We use our senses to find out about things.

- We use our senses to tell how things are the same or different.

- We group things that are the same.

Science Words

Match each word to a body part.

see   **hear**
feel   **smell**
taste

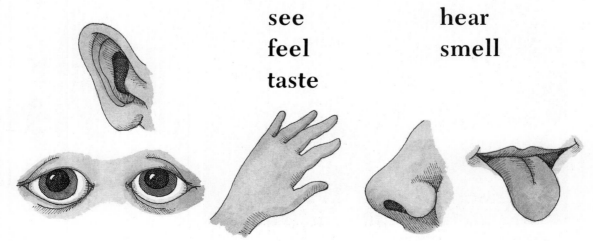

Questions

1. How are these the same?

2. How are they different?

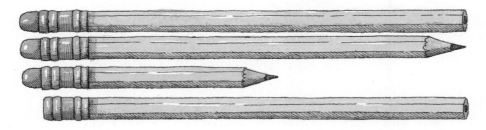

3. Which ones make a group?
   Why?

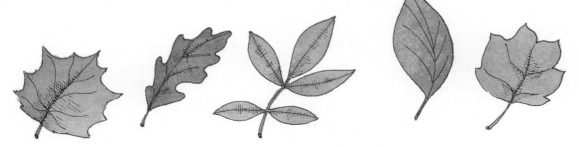

Science Project

Put things into a bag.
Let someone feel inside.
Can he or she guess what they are?

CHAPTER 2

LIVING THINGS

1.

IS IT ALIVE?

The rabbit is **alive**.
The toy rabbit is not alive.
Toys are not **living things**.

This plant is alive.
It is a living thing.

This plant is not alive.
What is it made of?

These things are not living.
How do you know?

Which things are living?
Which things are not living?

2.

ALL LIVING THINGS

All living things grow.
All living things need food.
They need water and air.
Food, water, and air help them grow.
What do you need to live?

Living things make more living things.
Plants make seeds.
Animals have babies.

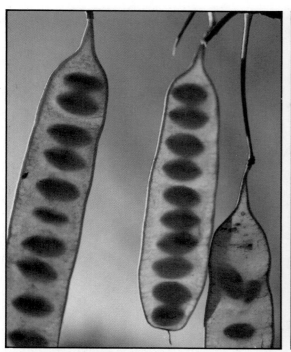

ACTIVITY

Are seeds living things?

1. Put some soil in a cup.

2. Put seeds in the soil.

3. Label your cup.

4. Water your seeds.

5. Do they grow?

6. Are they living things?

Name:_____
Kind of seed:_____
Date planted:_____
Date I saw leaves:
_____
Date I saw stem:
_____

3.

PLANT OR ANIMAL?

Plants are living things.
They grow.
They need food.
Green plants make their own food.
Plants do not move from place to place.

21

Animals are living things.
Animals move from place to place.
They move to find food.
Animals eat food.

Here are living things.
Which are plants?
Which are animals?
How do you know?

ACTIVITY

Is it a plant or an animal?

1. Find pictures of living things.

2. Put them into the right groups.

3. Find the name of the biggest living plant.

4. Find the name of the biggest living animal.

This is a house for green plants.
It is called a greenhouse.
The plants get sun here.
The workers give the plants water.
People come here to buy the plants.

Main Ideas

- Living things need food, water, and air.

- Living things grow.

- Living things make more living things.

- Plants and animals are living things.

Science Words

Which things are **living things?**

REVIEW

Questions

Look at these things.

1. Which ones can grow?

2. Which ones need water?

3. Which are plants?

4. Which are animals?

Science Project

Your food comes from living things.
Find pictures of foods.
Draw what they came from.

CHAPTER 3

AT HOME ON EARTH

1.

FORESTS

Plants and animals live on the **earth**.
Some parts of the earth are dry.
Some parts are rainy.
Some are windy.
Some are flat.

A **forest** is a home for living things.
A forest has many trees.
The trees need a lot of rain.

Trees here grow close together.
The trees give shade.
The forest feels cool.
Do you live near a forest?

ACTIVITY

Are all forest trees the same?

1. Collect leaves from many trees.

2. Make leaf rubbings.

3. Are the leaves the same size?

4. Are they the same shape?

2.

PLAINS

The **plains** are a home for living things.
The land of the plains is flat.
It does not get very much rain.
It is a windy place.

Grass grows on the plains.
Only a few trees grow here.
The plains are hot and dry in summer.

The plains are very cold in winter.

The plains have thick,
black soil.
Some animals dig in it.
This animal lives in
the soil.

Many plains animals
are big.
They eat the grass.
They have warm fur.
They can run very fast.

3.

DESERTS

A **desert** is a home for living things.
It is a dry place.
It hardly ever rains there.
Deserts can be flat or hilly.

The desert day is very hot.
The sun makes it hot.
Many desert animals hide
during the day.

Now it is night.
The animals come out to eat.

ACTIVITY

Do desert plants save water?

1. Look at a desert plant.

2. Cut open a leaf.

3. Squeeze the leaf.

4. What is inside?

5. What do desert plants do with rainwater?

The person talking works in a park.
He is a park ranger.
A park ranger knows about the land.
Rangers talk to people.
They tell them about the park land.

Main Ideas

- The earth's land is different from place to place.

- Forests, plains, and deserts are homes for living things.

- Some places get more rain than others.

Science Words

Match a word to each sentence.

forest   **plains**   **desert**

1. This home is hot and dry.

2. This home has many trees.

3. This home has a lot of grass.

Questions

1. Which place gets the most rain?

2. Which gets the least?

3. Do fish live in the desert?
How do you know?

Science Project

Make a picture of where you live.
Show the animals that live there.
Show the plants that live there.

CHAPTER 4

STAYING ALIVE

1.

MOVING

Animals **move** to find food.
Animals move to get away from danger.
This animal lives in the forest.
It jumps from tree to tree.

A fish lives in water.
It swims to move.
Where do these animals live?
How do they move in their homes?

Where do these animals live?
Do they move fast or slow?

Ducks live in water,
on land, and in the air.
They can move in three ways.
Do you know how?

2.

FINDING FOOD

Animals find food where they live.
Their bodies help them get the food.
This owl lives in the forest.
It has sharp claws to catch mice.

Food is hard to get.
What does each one eat?
Look at their bodies.
How do they get their food?

ACTIVITY

How do worms find food?

1. Put some soil in a jar.

2. Put in some worms.

3. Add some apple skin.

4. Make holes in the lid.
   Put it on the jar.

5. Wrap paper around the jar.
   Wait a few days.

6. Unwrap the paper.
   What do you see?

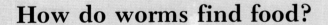

3.

STAYING SAFE

Animals have ways to stay safe.
This animal moves slowly.
It can not run away.
It has a hard **shell**.
The shell keeps it safe.

Colors help keep animals safe.
The rabbit is white.
The snow is white.
It is hard to find the rabbit in the snow.

Color keeps this animal safe.
What else keeps it safe?

Plants have ways to stay safe.
This plant has an **oil**.
The oil makes you itch.

These have sharp **thorns**.
Don't touch them!

ACTIVITY

Does color help animals?

1. Go outside. Put 4 sticks in the ground to make a square. Tie a string around the sticks.

2. What color is inside the square?

3. Cut out animal shapes. Use different colors.

4. Put the animals inside the square.

5. Which animals are easy to see?

PEOPLE AND SCIENCE

There are many animals in the zoo.
They eat many different things.
Zookeepers take care of the animals.
They know what the animals need.
These seals are eating fish.
How do they eat?

Main Ideas

- Animals move in different ways.

- Animals use their body parts to find food.

- Plants and animals have many ways to stay safe.

Science Words

Tell about the pictures.
Use these words.

shell　　　**oil**
colors　　**thorns**

REVIEW

Questions

1. How do these animals move?

2. How do they get food?

3. How do they stay safe?

CHAPTER 5

PLACE

1.
NEAR OR FAR?

The players are **far**.
Far things look small.
The people watching are **near**.
Near things look big.

What is near?
What is far?
How do you know?

The moon looks small.
It is far from earth.

The moon is really big.
How do you know?

2.

WHERE IS IT?

What a mess!
Toys are **under** the bed.
Socks are **over** the lamp.
The dog is **on** the bed.

Hide and seek!
The boy is **in front** of the rock.
The girl is **behind** the rock.
Who is near?
Who is far?

ACTIVITY

What is near or far?

1. Go outside.
   Look in front of you.

2. What is near?
   What is far?

3. Write it down.

3.

DID IT MOVE?

The bus is far away.
It looks small.

The bus is nearer.
It looks bigger.
The bus moved.

63

Here comes a train!
The train is moving.

Moving things change places.
What things did not move?

The rocket is very big.
It is near the ground.

Blast off!

The rocket is moving.
How do you know?

ACTIVITY

What moved?

1. Cut out some shapes.

2. Put them in a row.

3. Tell a friend to close his or her eyes.

4. Move one shape.

5. Ask which one was moved.

6. How does your friend know?

PEOPLE AND SCIENCE

This person tells cars when to move.
Some cars are stopped.
Some cars are moving.

CHAPTER

Main Ideas

- You can tell if things are near or far.

- We use words to tell where things are.

- You can tell if things have moved.

Science Words

Use your science words.
Tell where things are.

| over | in | behind |
| under | on | in front |

REVIEW

Questions

1. What is near?

2. What is far?

3. What moved?

4. What did not move?

CHAPTER 6

TIME

1.

MORE OR LESS TIME?

You can make a toy
house.
It takes a **short** time.

This house takes a **long** time.
A real house takes **more** time to make.

This is a race.
One skater goes fastest.
It takes him **less** time.
He wins the race.

ACTIVITY

Can sand help time things?

1. Make a hole in the bottom of a cup.

2. Hold the cup over a plate.

3. Fill the cup with sand.

4. What happens?

5. Have a race with a friend.
   Write your name 10 times.

6. Use the sand timer to find
   out who takes less time.

2.

ONE DAY

A new **day** begins.

The sun seems to come up.
Part of a day is light.

The sun seems to go down each day.

Part of a day is dark.
What makes a day light and dark?

We eat, work, and sleep.
All these things take one day.

ACTIVITY

What do you do in a day?

1. Use 3 sheets of paper.

2. Write *morning* on one sheet.

3. Write *afternoon* and *night* on the other two.

4. Write down what you do for one day.

3.

DAY AFTER DAY

Some things take more than one day.
It takes many days for a puppy to grow up.

It takes three days to make this jelly.

Yesterday.

Today.

Tomorrow.

What will the field look like tomorrow?

PEOPLE AND SCIENCE

Many people are in this race.
How much time does each one take?
There is a clock at the end.
The clock tells how much time has passed.

Main Ideas

- Some things take more time to do than others.

- One day has a daytime and a nighttime.

- Some things take many days to do.

Science Words

Use these words.
Tell about the picture.

short   **more**   **day**
long   **less**

REVIEW

Questions

1. What is the right order?

2. Yesterday, today, and tomorrow.
Which is which?

Science Project

Make a book.
Use one page for a day.
Draw what you do each day.

CHAPTER 7

HOW BIG IS IT?

1.

SIZE

Some hats are **big**.
Some hats are **small**.
You can tell **size** by looking.
You can tell size by feeling.

You can **measure** size.

Which lid fits?
You can try each size.

The man measures the girl's foot.

Then he knows what size shoe she needs.

Does it fit?
How do you know?

How much do you want?
The cups help measure.
There are three sizes.
Pick one!

2.

HOW LONG?

Look at things side by side.
You can tell what size they are.
The clown's shoe is **long**.
The girl's shoe is **short**.

We need a big tank.
Which one is longer?
Use two strings to find out.

Put the strings next to each other.
Which tank is longer?

ACTIVITY

How long is it?

1. Hook paper clips together.

2. Lay them next to your pencil.

3. How many paper clips did you use?

4. Measure other objects.

5. Which object is longest?
   Which one is shortest?

3.

HOW HEAVY?

Big pumpkins are **heavy**.
Small ones are **light**.
The boy feels how heavy they are.

The teacher is heavier.
She **weighs** more.

The girl is lighter.
She weighs less.

The teacher can lift the girl.

We measure how heavy things are.

These people weigh tomatoes.

They use a **scale**.

This is a scale, too.

ACTIVITY

How much does it weigh?

1. Put something on a scale.

2. Look at the number it points to.

3. Write the number on a chart.

4. Weigh other things.

5. Which one is lightest?

6. Which one is heaviest?

Have you ever seen this sign?
It tells a truck driver to stop.
The truck will be weighed.
Trucks that are too heavy can not go on.
Heavy trucks can break the road.

Main Ideas

- Things are different sizes.

- We measure how long things are.

- We measure how heavy things are.

Science Words

Match a set of words to each picture.

> **big—small**
> **long—short**
> **heavy—light**

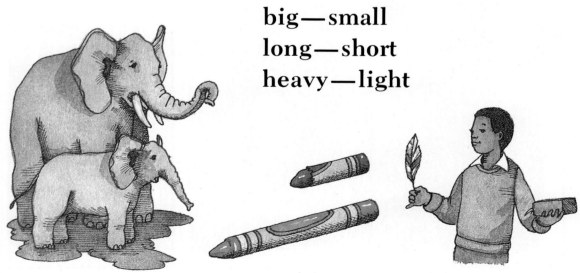

REVIEW

Questions

1. Is this your size?
How can you tell?

2. What does each one measure?

Science Project

Make a folding book.
Draw what you wear.
Tell about the sizes.

CHAPTER 8

PEOPLE GROW AND CHANGE

1.

PEOPLE ARE LIVING THINGS

You are a living thing.
You can move.
You move from place to place.
How are people like other animals?

People need food to live and grow.
Our foods come from animals.
Our foods come from plants.

People have babies.
The babies will grow up.
They need food, water, and air.

2.

WE GROW

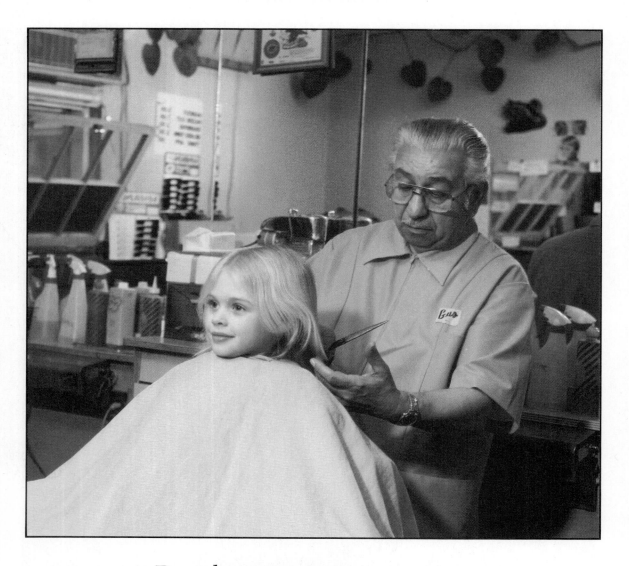

People **grow** in many ways.
Your hair grows.
Your nails grow.
You get bigger.

You grow taller.
You grow heavier.
Growing up takes many years.
You will stop growing one day.
Then you will be an **adult**.

We are not all the same size.
Some people grow faster than others.

Look at your baby picture.
How have you grown?

ACTIVITY

How do you grow?

1. Measure your weight.
2. How much do you weigh?
3. Measure your size.
4. How tall are you?

3.

WE CHANGE

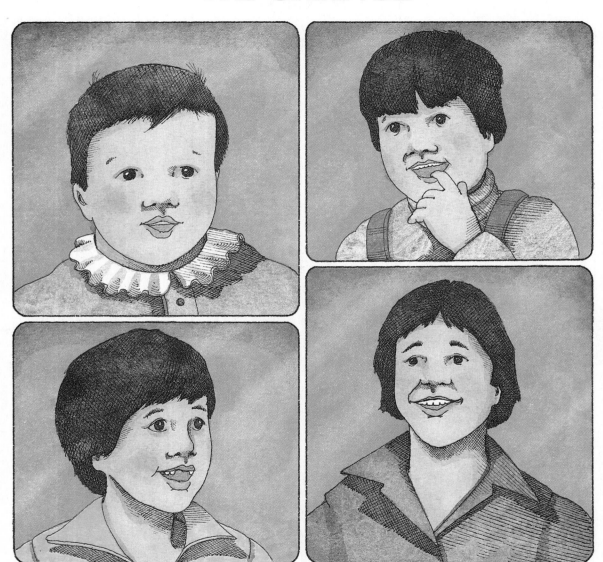

People **change** as they grow.
Your **teeth** change as you grow.
People have two kinds of teeth.

Children grow bigger.
Adults change in other ways.
Look at their bodies.
What changes can you see?

ACTIVITY

How have you changed?

1. Bring your baby pictures to school.

2. How has your weight changed?

3. How else have you changed?

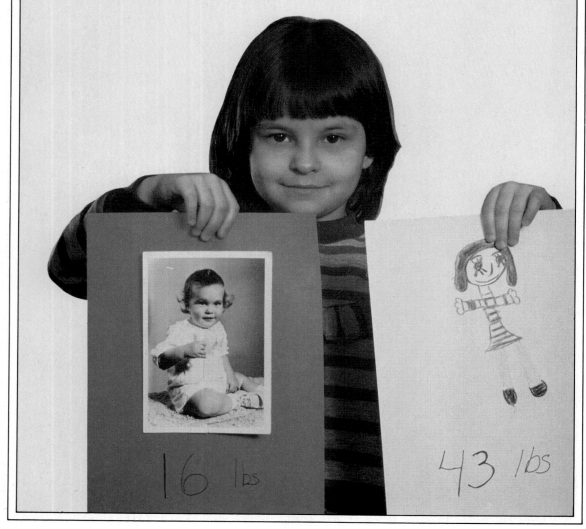

16 lbs

43 lbs

PEOPLE AND SCIENCE

This nurse works in a school.
She checks how children grow.
She measures the children.

Main Ideas

- People are living things.

- Children grow until they are adults.

- Your body changes as you get older.

Science Words

Tell about the picture.
Use these words.

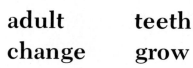

adult   **teeth**
change   **grow**

Questions

1. These are living things.
   How are they the same?

2. How do people grow and
   change?
   Show the right order.

3. What can the boy do?
   What can a baby do?
   How is it different?

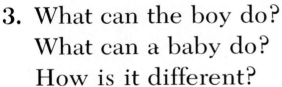

CHAPTER 9

MAGNETS

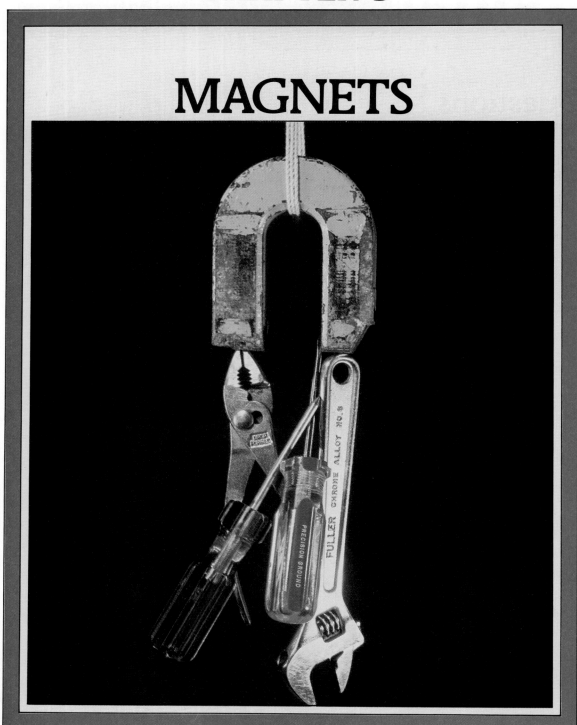

1.

MAGNETS CAN PULL

A **magnet** can pull a tack.
A tack has **iron** in it.
Magnets pull things made of iron.

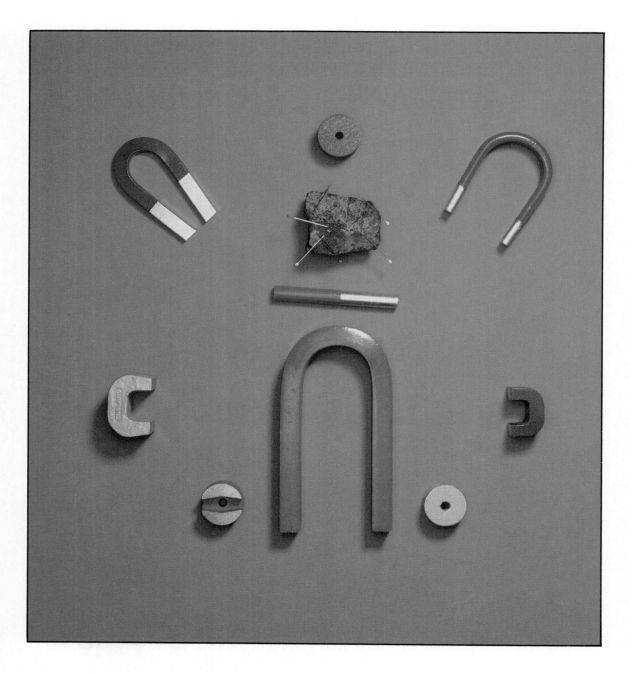

Magnets are made of iron, too.
Some magnets are made by people.
Some are found in rocks.
Magnets are many shapes and sizes.

ACTIVITY

What things can magnets pull?

1. Touch a magnet to many things.

2. What things does the magnet pull?

3. What things does it not pull?

4. How are the things it pulls the same?

2.

MAGNETS HAVE POLES

Magnets have two ends.
They are called **poles**.
One pole is marked N.
One pole is marked S.

The boy puts two poles together.
The two poles are alike.
He feels a **push**.
Poles that are alike push away.

The girl puts two poles together.
The poles are different.
She feels a **pull**.
Poles that are different pull together.

118

ACTIVITY

What part of a magnet is strongest?

1. Hang clips from a magnet one by one.

2. Hang them from the N pole.

3. Now, hang them from the S pole.

4. Try the middle of the magnet.

5. Which part holds the most clips?

3.

PEOPLE USE MAGNETS

We use magnets every day.
Magnets hold things in place.

This toy has a magnet.
The magnet pulls the bits of iron.
The magnet can pull through the plastic.

ACTIVITY

You can make a magnet.

1. Hold an iron nail in one hand.
   Hold a magnet in the other.

2. Push the magnet across the nail.
   Do this 10 times.

3. Try to pick up a clip with the nail.

4. Is your new magnet strong?

PEOPLE AND SCIENCE

The boy holds a compass.
It has a magnet inside.
The magnet always points
to the north.
The compass helps the boys
find their way.

Main Ideas

- Magnets pull on things made of iron.

- Every magnet has two poles.

- People use magnets in many ways.

Science Words

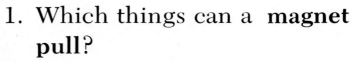
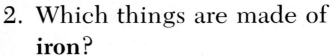

1. Which things can a **magnet pull**?
2. Which things are made of **iron**?

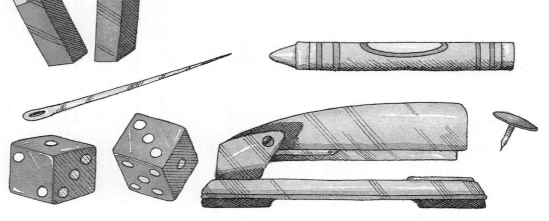

Questions

1. Look at the magnets.
Where are their poles?

2. Would you feel a push or
a pull here?

Science Project

Can a magnet work through
things?
Find out.
Use paper clips and cardboard.

CHAPTER 10

AIR

1.

AIR IS REAL

Air is all around us.

We can not see, smell, or taste air.

We can see things moved by air.

We can smell things in air.

We can feel air.
We can hear air when it moves.
Moving air is called **wind**.

We use air in many ways.
How can we use air?

Living things need air.
We breathe air.
Clean air is good for us.
Dirty air is not good for us.

ACTIVITY

Can you make air move things?

1. Make a fan.

2. Use the fan to make wind.

3. Can you hear the air?
   Can you feel it?
   Can you see it?

4. What can you move with your fan?

2.

AIR FILLS SPACES

The boy fills the bag with air.
Air fills the **space** in the bag.

Air is almost everywhere.
There is air in this glass.

Air has no shape.
It can take any shape.
Air fills the spaces of things.

ACTIVITY

Does air take up space?

1. Put paper in a cup.

2. Turn the cup upside down.

3. Put it in water.

4. Take the cup out of the water.

5. Feel the paper.

6. Is it wet or dry?

7. Where was the air?

3.

AIR HAS WEIGHT

Air can push things.
The boy put a bag under the bear.
He blew air into the bag.
What made the bear move?

We can weigh air.

How much does this ball weigh?

The air has been pushed out of the ball.

How much does it weigh now?

Which time was it heavier?

ACTIVITY

Can you weigh air?

1. Tie a string to the middle of a stick. Your teacher will hang the stick from the ceiling.

2. Tie a string to each end of the stick.

3. Blow up a balloon. Tie it so that the air stays inside.

4. Blow up another balloon only half way.
Tie it shut.

5. Tie a balloon to each string.

6. Which balloon has more air?

7. Which balloon weighs more?

PEOPLE AND SCIENCE

There is no air in outer space.
This person wears a special suit.
There is air inside the suit.

CHAPTER

Main Ideas

- Air is all around us.

- Air can push things.

- Air takes up space.

- Air has weight.

Science Words

Tell about the picture.
Use these words.

space   **wind**   **air**

REVIEW

Questions

1. Which senses tell you about air?

2. Which jars have air in them?

Science Project

How dirty is the air?
Put some jelly on a card.
Put the card outside for a day.
What is on the jelly?

CHAPTER 11

ROCKS

1.

THE EARTH'S ROCKS

The earth is made of **rocks**.
You can see some of the rocks.

Most rocks are under the ground.
This cave is under the ground.
How do the rocks look?

You can find rocks near water.
Water changes the rocks.
Water breaks rocks.
Sand is made of very small rocks.

2.

ROCKS ARE DIFFERENT

Rocks look different.
They are different colors.
Some are **shiny**.
Some are **dull**.

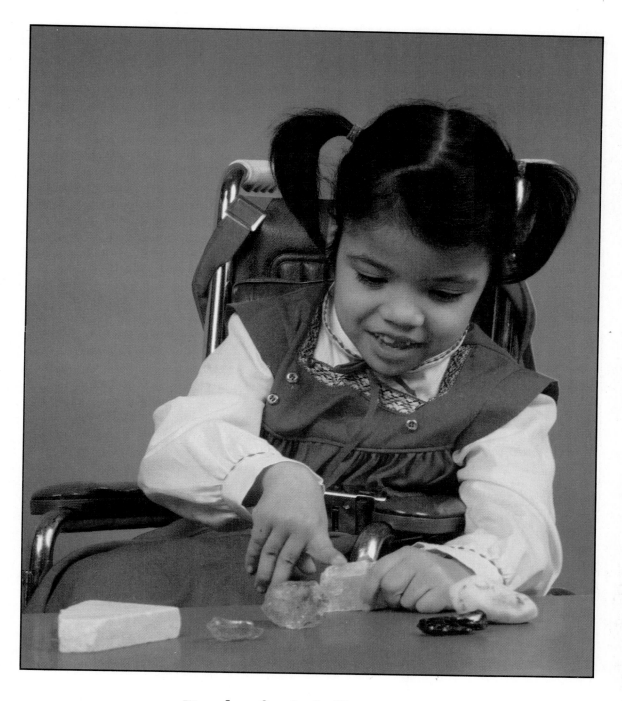

Rocks feel different.
Some are **rough**.
Some are **smooth**.

You can rub rocks together.
Hard rocks are strong.
Soft rocks fall apart.

ACTIVITY

How hard are rocks?

1. Scratch a rock with a nail.

2. Do bits break off?

3. Try scratching many rocks.

4. Which ones are soft?

5. Which ones are hard?

3.

PEOPLE USE ROCKS

People use rocks.
They cut them from the earth.
The rocks are carried away.

People build with rocks.
What other things are rocks used for?

ACTIVITY

Can you group rocks?

1. Bring some rocks to school.

2. Put the rocks in groups.

3. Give each group a name.

PEOPLE AND SCIENCE

These people are digging for gems.
They are hard to find.
Gems are pretty rocks.
People like to wear gems.

Main Ideas

- The earth is made of rocks.

- Rocks look and feel different.

- Rocks can be grouped.

- People use rocks in many ways.

Science Words

Match the opposite words.

shiny	soft
rough	dull
hard	smooth

Questions

1. Where can you see these
rocks?

2. Chalk is a soft rock.
How can you tell?

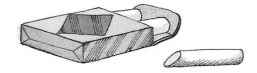

3. How can you use this rock?

CHAPTER 12

SOIL

1.
WHAT IS SOIL?

The land has **soil** on it.
You walk on soil.
We build on soil.
People dig in the soil.

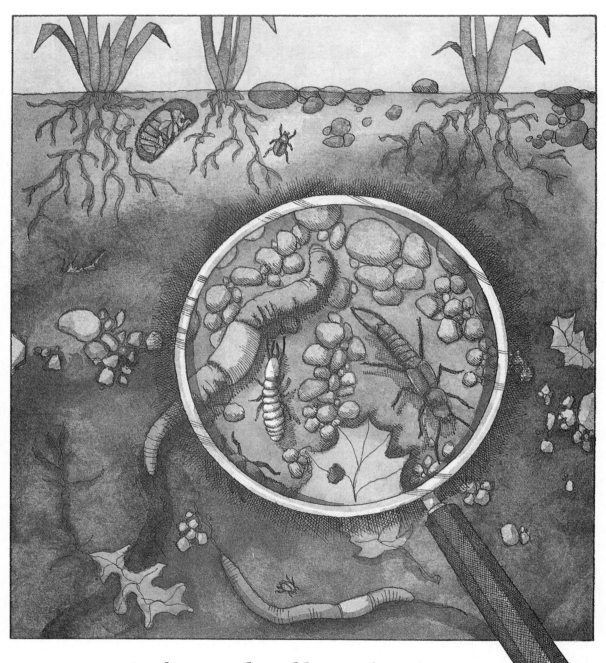

Soil is made of bits of rock.
There is air and water in soil.
Soil has dead plants and
animals in it.

Soils are different.
They are made of different rocks.
They look different.
They feel different.

ACTIVITY

What is in soil?

1. Look at soil from
   under a bush.
   Use a hand lens.

SAND	SOIL FROM BUSH	SOIL FROM FIELD

2. Look at sand.

3. Look at soil from a field.

4. Write down what you see.

2.

LIVING THINGS NEED SOIL

Plants grow in soil.
They take in water from the soil.
They take in air from the soil.
They use tiny bits of rock.

Animals need soil.
Some live in the soil.
Some make their homes
with soil.

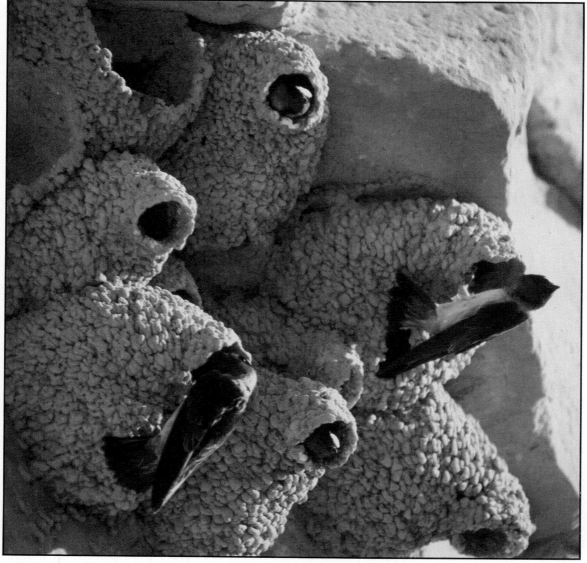

People need soil.
We grow plants for food and clothes.

We grow trees for wood.

ACTIVITY

Is there air in soil?

1. Put some soil in a jar.

2. Pour water on it.

3. What comes out of the soil?

PEOPLE AND SCIENCE

This house is built in a hill.
Soil is on the roof.
Grass grows on the roof!

Main Ideas

- Soil is made of rocks, water, air, and dead plants and animals.

- Living things need soil.

- People use soil in many ways.

Science Words

What things is **soil** made of?

REVIEW

Questions

1. What do plants get from soil?

2. How do people use soil?

Science Project

Get soil from three places.
How are the soils the same?
How are they different?

SCIENCE WORD LIST

The science words are in ABC order. The number tells you on which page to find the word.

adult	103	forest	30
air	127	group	8
alive	15	grow	102
animals	22		
		hard	146
behind	61	hear	2
big	85	heavy	91
change	106	in front	61
colors	50	iron	113
day	74	less	72
desert	36	light	91
different	5	living things	15
dull	144	long	71, 88
earth	29	magnet	113
		measure	86
far	57	more	71
feel	1		

Illustration by Lane Yerkes

HRW Photos by Bruce Buck appear on pages: 20, 77, 135 (both), 136.

HRW Photos by Russell Dian appear on pages: 21, 61, 64 Courtesy of FAO Schwartz, 74, 75, 80, 87, 88, 101, 121, 134, 145, 147, 150.

HRW Photos by Ken Lax appear on pages: 17, *bottom* 45, 58, 64, 79, 86, 92, 93, 104, 105, *left* 128, 131, 146.

HRW Photos by Richard Hutchings appear on pages 2, 7.

HRW Photos by Yoav Levy/Phototake appear on pages: 3, 4, 5, 8, 9, 10, 16, 24, 34, 38, 48, 66, *top right* 71, 73, 89 Courtesy of The Norwalk Aquarium, 90, 94, 102, 108, 109, 113, 115, 117, 118, 119, 122, 130, 133, 157, 162.

HRW Photo by David Lokey/Vail Associates appears on page 76.

HRW Photos by Ken Karp appear on pages 85, 144.

HRW Photos by Louis Fernandez appear on pages 15, 60, 63, 95, 103, 120.

HRW Photos by Richard Haynes appear on pages 114, 116, 132, 158.

HRW Photos by Carole Lynch appear on pages 52, 62.

HRW Photo by John Running appears on page 51.

HRW Photo by Michael Provost appears on page 6.

HRW Photo by Russ Kinne appears on page 141.

TOC Credits

Chapters 1–3: p. 4 — © George Hausman/The Image Bank. Ralph Breswitz/DPI; Charlie Ott/DPI
Chapters 4–6: p. 5 — © Russ Kinne/Photo Researchers, Inc.; Nicole Bengiveno/Marriott's Great America; Manuel/The Image Bank
Chapters 7–9: p. 6 — © Bastcott/Momatiuk — Photo Researchers, Inc.; © 1981 Brownie Harris; Runk/Schoenberger — Grant Heilman
Chapters 10–12: p. 7 — John Desjardins/DPI; © Judy Griesdieck/Leo de Wys, Inc.; Grant Heilman

Chapters 1–6: p. 0 — George Hausman/The Image Bank; p. 1 — Hal McKusick/DPI; p. 11 — Syd Greenberg/DPI; p. 14 — Ralph Breswitz/DPI; p. 15 *top* — Leonard Lee Rue III/Animals, Animals; p. 18 — Grant Heilman; p. 19 *top* — Helen Cruicshank/N.A.S./Photo Researchers, Inc., *bottom* — Leo de Wys, Inc.; p. 22 *top* — J. Alex Langley/DPI, *bottom* — Alon Reininger/DPI; p. 23 *bottom right* — © Kjell Sandred/Photo Researchers, Inc., *top right* — Leonard Lee Rue III/Monkmeyer, *top left* — Phil Dotson/DPI, *bottom left* — Ronald Thomas/Taurus; p. 25 — Michal Heron/Monkmeyer; p. 28 — Charlie Ott/DPI; p. 29 — Tom McHugh/Photo Researchers, Inc.; p. 30 — Mike Howell/Leo de Wys, Inc.; p. 31 — W.H. Hodge © Peter Arnold, Inc.; p. 33 — Everett Johnson/Leo de Wys, Inc.; p. 34 — *top* John Zoiner/Peter Arnold, Inc., *bottom* Grant Heilman; p. 35 *top* — © David Halpern/Photo Researchers, Inc., *bottom* — Leonard Lee Rue III/DPI; p. 36 — © David Hiser/E.P.A. Documerica; p. 37 *top* — Peter B. Kaplan/Photo Researchers, Inc., *bottom* — Panuska/DPI; p. 39 — Mimi Forsyth/Monkmeyer; p. 42 — Russ Kinne/Photo Researchers, Inc.; p. 43 — Tom McHugh/Photo Researchers, Inc.; p. 44 *top left* — Steinhart Aquarium/Photo Researchers, Inc.; *bottom left* — © Vance Henry/Taurus, *bottom right* — Martin Vanderwall/Leo de Wys, Inc., *top right* — © Tom McHugh/Photo Researchers, Inc.; p. 45 *top left* — M.P. Kahl/Bruce Coleman, Inc., *top right* — David Overcash/Bruce Coleman, Inc.; p. 46 — Phil Dotson/DPI; p. 47 *top right* — James Karales/Peter Arnold, Inc., *bottom right* — Herbert Lanks/Monkmeyer, *left* — Philip Kahl, Jr./Photo Researchers, Inc.; p. 49 — Leonard Lee Rue III/Monkmeyer, *inset* — W.M. Partington/N.A.S./Photo Researchers, Inc.; p. 50 *top* — Leonard Lee Rue III/Bruce Coleman, Inc., *bottom* — Animals, Animals; p. 51 *top* — Everett Johnson/Leo de Wys, Inc., *bottom right* — Syd Greenberg/Photo Researchers, Inc.; p. 53 — Leonard Harris/Leo de Wys, Inc.; p. 56 — Nicole Bengiveno/Marriott's Great America; p. 57 — Mickey Palmer/DPI; p. 59 *top* — ProPix/Monkmeyer, *bottom* — NASA; p. 65 *top left* — © 1981 Gerald Davis/Woodfin Camp & Assocs., *top right* — NASA/Woodfin Camp & Assocs., *bottom* — Frank Fournier/Woodfin Camp & Assocs.; p. 67 — Karen Collidge/Taurus.

Chapters 6–12: p. 70 — Manuel/The Image Bank; p. 71 — Ted Mahieu/The Image Bank; p. 72 — Alvis Upitis/The Image Bank; p. 78 — Ken Stepnell/Taurus; p. 81 *top* — Janeart Ltd./The Image Bank, *bottom* — Gerhard Gscheidle/Peter Arnold, Inc.; p. 84 — Bastcott Momatiuk/Photo Researchers, Inc.; p. 91 — Larry Dale Gordon/The Image Bank; p. 98 — © 1981 Brownie Harris; p. 99 — Jack Fields/Photo Researchers, Inc.; p. 100 *top* — Jack and Ray Ellena/DPI, *bottom* — Grant Heilman; p. 107 — Leo de Wys, Inc.; p. 112 — Runk/Schoenberger/Grant Heilman; p. 123 — Scott Witte; p. 126 — John Desjardins/DPI; p. 127 — © Eric Kroll/Taurus; p. 128 *bottom* — Steve Lissau, *top right* — Courtesy of Boeing; p. 129 *top* — F.C. Wilcox/DPI, *bottom* — Carter Hamilton/DPI; p. 137 — NASA; p. 140 — Judy Griesdieck/Leo de Wys, Inc.; p. 142 — David Hiser/The Image Bank; p. 143 — Charlie Ott/DPI; p. 145 — Eunice Harris/Photo Researchers, Inc.; p. 148 — Jim Steenson/Leo de Wys, Inc.; p. 149 *top right* — © Norman Thompson/Taurus, *bottom left* — Hugh Rogers/Monkmeyer, *top left* — Steve Niedorf/The Image Bank, *bottom right* — Wil Blanche/DPI; p. 151 — © Bruce Roberts/Photo Researchers, Inc.; p. 151 — Diamond Information Center; p. 154 — Grant Heilman; p. 155 — Hugh Rogers/Monkmeyer; p. 159 — Grant Heilman; p. 160 — © Allan Morgan/Peter Arnold, Inc., *inset* — © M.P. Kahl/Photo Researchers, Inc.; p. 161 *top* — © Larry Smith/DPI, *bottom* — Porterfield-Chickering/Photo Researchers, Inc.; p. 163 — © Randa Bishop/DPI.